This Isometric Graph Paper book belongs to:

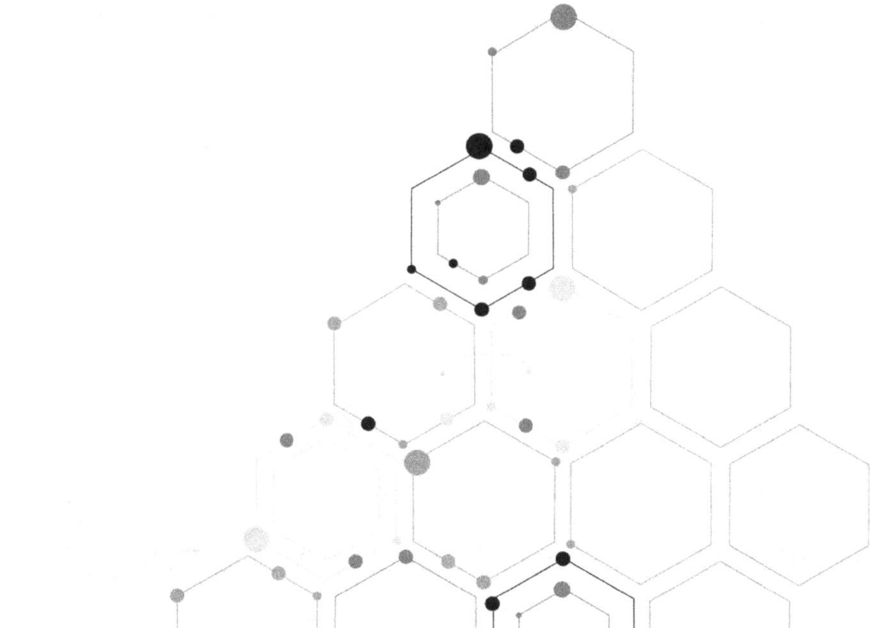

www.ingramcontent.com/pod-product-compliance
Lightning Source LLC
Chambersburg PA
CBHW081731220526
45468CB00008B/2052
9781723840968